CHESAPEAKE AND OHIO CANAL
Old Picture Album

By Thomas F. Hahn, I.A.

Published by
The American Canal & Transportation Center
P.O. Box 842, Shepherdstown, W.Va. 25443

Second Printing June 1977

Preface

For many years I have enjoyed studying the history, the engineering and the architecture of the Chesapeake and Ohio Canal. My work as Supervisor of the Restored Section of the C. & O. Canal and as a consultant for Industrial Archeology for the National Park Service, during which time Joseph R. Prentice and I worked nearly three years in the field on the canal compiling some 30 volumes of data, intensified that enjoyment. In this study we found that one of the best sources of information was historical photographs.

We begin to see photographs of the canal in the 1860s during the Civil War. In fact, some of the Brady/Handy photographs of that period remain as examples of excellence in photography today. After the war there is a paucity of photos until the late 1880s. Photographs then become more common and we have a fairly good selection in the first two decades or so through the end of the operating period in 1924. Unfortunately, photographers take pictures of the same things, so that there are many photos of the same places in the historic operating period, but many gaps. Beginning about 1938, with the acquisition of the canal by the National Park Service, we see a more systematic study of the canal as different considerations began to develop as to the use of the canal. For this reason, I have selected some photos which are historical, but not of the canal operating period, so as to lend a feeling of continuity as we travel from one end of the canal to the other.

My associate, Bill Shank, and I thought that with the American Bicentennial occurring this year, and with the 150th anniversary of the beginning of construction of the C. & O. Canal only two years away, that this would be the ideal time to bring out a book on some of the old time scenes on the old C. & O. Canal. I have selected 100 of some of my favorites to share with you. I hope you like looking at them as much as I did selecting them.

Captain Tom Hahn, USN (Ret.), I.A.

January 1976
Shepherdstown, West Virginia

A Short History of the Chesapeake and Ohio Canal

The Chesapeake and Ohio Canal was begun as a proposed waterway along the established Potomac and trans-Allegheny trade route to the Ohio River. It was the economic heir of the old Potomack Company which had attempted unsuccessively from 1785 to 1828 to fulfill George Washington's dream of developing a practical water route through the Potomac Valley.

Construction of the C. & O. Canal started on July 4, 1828. After much glowing oratory President John Quincy Adams turned the first spade of dirt. On the same day, construction of the Baltimore and Ohio Railroad began in Baltimore. The race for the West across the Alleghenies was on!

From the beginning the canal was beset with many problems. Lumber, building stone, lime for cement, and other building supplies were often scarce and expensive. Excavation frequently revealed unexpected hardpan, slate, or gravel which made the work difficult and costly. Many difficulties were encountered in securing land titles for the right-of-way. Land costs were frequently above estimate and excessively high.

Because the Potomac was predominantly agricultural, skilled labor was virtually unobtainable. The canal company turned to Europe for indentured labor. Men arriving from Ireland, Germany, Netherlands, England and Wales soon provided a pool of stone cutters, masons, carpenters, and laborers. Welsh miners were imported to dig Paw Paw Tunnel. Many workers became the unwitting victims of unscrupulous contractors. Food was often poor and housing consisted of makeshift dwellings. Diseases swept through the labor camps. Because of rivalries and prejudices the workers brought with them from Europe, there was much fighting between groups. Sometimes the men rioted when they weren't paid and the militia had to be called out. On top of these troubles, the B. & O. Railroad continued to press westward.

Canal and railroad were soon competing for property rights. The C. & O. Canal Company had decided to locate its waterway along the north bank of the Potomac River in Maryland. The B. & O. was organized in such haste that the final decision concerning its route was not made until after it had gotten underway. The railroad then decided to construct its line from Baltimore through Frederick to Point of Rocks in the Potomac Valley. Confident of securing the right-of-way in the Potomac Valley inherited from the old Potomack Company, the C. & O. Canal Company directors proceeded in a leisurely fashion to secure necessary land titles. However, the railroad moved up aggressively and sent agents far up the valley to secure land rights from the inhabitants at many points – particularly at the narrow gorge above Point of Rocks. Because the bluffs came close to the river, there was room for only one of the lines. The B. & O. Railroad claimed the right-of-way and secured an injunction against the canal company. The canal company appealed and the courts eventually ruled in the canal's favor – thus ending a four-year bitter struggle.

The canal was used as each section was completed: first, from Georgetown to Seneca in 1831; then to Harpers Ferry in 1833; and to near Hancock in 1839, 134 miles from its beginning in Georgetown. About this time the canal encountered serious financial problems and the next 50.5 miles to Cumberland were not completed until 1850 – eight years after the railroad reached that point. And with that, the C. & O. Canal Company dropped its plans to continue another 180 miles.

The company had spent $12 million to build 184.5 miles of canal consisting of 74 lift locks which raised it from sea level at Georgetown to 610 feet at Cumberland, 11 stone aqueducts which carried the canal over major Potomac tributaries, seven dams to supply water for the canal, a number of waste weirs to control the water level, about 200 culverts to carry roads and streams under the canal, a 3,118-foot tunnel to take it under a mountain, and an assortment of stop gates, river locks, guard locks, bridges, shops, section houses, and lock houses.

The canal generally was 50 to 60 feet wide at towpath level, sloped to 30 to 40 feet at the bottom and carried a minimum depth of six feet of water. The size of the locks – 15 feet wide and 90 feet long in the clear – restricted the size of the boats. A typical canal boat was 14.5 feet wide, 92 feet long (by putting the rudder to the side it could fit in the 90 foot lock), drew 4.5 feet of water, and could carry up to 135 tons of cargo, though 120 tons of coal was the normal maximum load.

The canal did not attain any great measure of economic success, but it provided a prominent and leisurely means of transporting coal, flour, grain, and lumber to Washington. In the early summer of 1889, a titanic flood swept the Potomac Valley, leaving the canal in ruin. However, it was rebuilt and used until 1924, when another flood seriously damaged the already financially troubled canal company.

The canal and its property were sold by the receivers of the canal company in 1938 to the United States, under the stewardship of the National Park Service of the Department of Interior. During the period 1939-1942 the lower 22 miles of the canal were restored and watered. In 1961, within hours of President Kennedy's inauguration, President Eisenhower declared the canal a National Monument. After many years of unsuccessful legislation, the Canal finally became the Chesapeake and Ohio Canal National Historical Park on 8 January 1972.

Unfortunately, just at the time restoration began to get underway, along came Tropical Storm Agnes in June of 1972, severely damaging many of the fragile structures of the canal. Since that time, funds appropriated by Congress have been spent mostly on the preservation of weakened aqueducts, culverts and locks, with much of the towpath restored to provide continuity for hikers, bikers and horseback riders. *(Adapted from the interpretive brochure on the C. & O. Canal by the National Park Service.)*

Lock House B on the Washington Branch of the C. & O. Canal. Its location at 17th & Constitution in Washington, D.C. is at its original site. This branch canal, connecting the main line of the C. & O. at its Georgetown terminus at the mouth of Rock Creek, brought cargo into the markets of Washington, and formed a link with the Washington City Canal. (Library of Congress, Brady/Handy photo, c. 1860)

Georgetown Tidal Lock at the confluence of Rock Creek and the Potomac River. This is "Mile Zero" of the C. & O. Canal. The structure near the center is a tumbling dam, with lift gates, across Rock Creek. After passing through the Tidal Lock, boats traveled .3 miles up Rock Creek to Lift Lock 1, where they entered the canal proper. The beginning of the Washington Branch of the C. & O. would have been about where the gas tank is on the far side of Rock Creek.

Rock Creek at 25th and Pennsylvania. Canal boats used Rock Creek from the Tidal Lock to near this point where they entered Lock 1 in Georgetown. In the foreground is an unidentified packet. On the bridge (carrying Washington's water supply as well as vehicular and pedestrian traffic) can be seen trolley cars of the Capitol Traction Company.

Lock 1, Georgetown. The City of Washington can be seen in the background. Though canal boats entered Rock Creek at the Tidal Lock .3 mile below, this is the beginning of the cut canal itself. The towpath is normally on the river side of the canal, but in Georgetown, between Locks 41 and 42 and from Lock 45 to Lock 46 it is on the land side of the canal.

Lock 3, Georgetown. The "Veterinary Hospital" across the lock previously housed the Duvall Foundry which was built c. 1856. The towpath was put on the land side of the canal in Georgetown so as to keep the commercial area on the river side free for a road and for loading and unloading. The freight boat passing through the lock is "light" or unloaded as indicated by the way it rides high in the water. Coal was the normal cargo from Cumberland downstream; boats traveled with little or no cargo on the return trip upstream. (National Park Service)

Wisconsin Avenue (Old High Street) Bridge was one of five stone bridges over the canal in Georgetown. Faded plaques in the face of the Aquia freestone list C. H. Dibble as Builder, Charles Mercer as President (of the canal company), Thomas Purcell Superintendent, Andrew Jackson President. (Library of Congress)

Francis Scott Key lived in this house near present-day Key Bridge. In this house, Key is said to have first read the Star Spangled Banner before a meeting of his glee club. The Key family left the house in the 1830's because of the turbulence caused by the construction of the canal in back of their house.

The Potomac Aqueduct, with Georgetown across the river. The arches in the abutment on the Maryland side of the river were to accommodate the Baltimore and Ohio Railroad. The figures in the foreground are typical of those found in Brady/Handy photographs. The Potomac Aqueduct, built 1833-1843, carried the 7 mile 416 yard Alexandria Canal on its way from its connection with the C. & O. Canal in Georgetown to Alexandria. (Library of Congress, 1865)

After the Civil War in 1868, a new truss was built to support the Potomac Aqueduct and it was reinforced by huge side arches. A road was added on top of the trusses, making a two-level bridge with canal traffic on the bottom and road traffic on top. (Library of Congress, c. 1879-1887)

This is a very old drawing of the western end of Georgetown looking west along what was later Canal Road (or Aqueduct Road) from what is now Key Bridge. The canal later flowed through this same area between the road and the river, with the same Three Sister Islands in view in the river. (Library of Congress)

Canal boats stopped above Georgetown to await orders as to where they were to unload. In 1875, the peak year of the canal, nearly one million tons of cargo were carried in about 540 boats, typical of those tied up here along the towpath. (Thompson Collection, National Park Service)

The Ludlow Patton, steam canal boat, made its maiden run from Cumberland to Georgetown in December 1875 in record time. The patent for its propeller-raising device was granted to Henry G. Wagner of Georgetown who built the boat, on 21 April 1876. The boat was built in Cumberland, but the machinery was made at the Duvall Foundry in Georgetown. (The National Gallery of Art, 1875)

The Mission on the Towpath was established first in 1894 by Mrs. S. E. Safford, in a small brick house, then in 1913 in this frame building. Mrs. Safford taught canallers' children to read and write, instructed adults in the industrial arts to help them find jobs off the canal in the winter and assisted them in finding housing in the winter.

The Georgetown Inclined Plane, completed in 1876, carried boats from canal level to river level via wooden caissons mounted on wheels. It was operated initially by water power and then by steam, remaining in operation several years. Once at the bottom of the incline, boats made their way into the river via the outlet shown here. (Thompson Collection, National Park Service)

Chain Bridge over the canal and Potomac River during the Civil War was one of the three bridge crossings from the District of Columbia into Virginia. The first Union Army sentinel to be court-martialed for sleeping at his post during the war had fallen asleep at the District end of the bridge, but he was later saved from execution by President Lincoln. (William M. Smith Photo, Library of Congress c. 1862)

Lock 6 at Brookmont was called Magazine Lock, (so named for the U.S. Powder Magazine located just around the curve of the canal, near the spot where President John Quincy Adams turned the first spade of dirt for the beginning of the C. & O. Canal on 4 July 1828.) The locktender is about to lower the drop gate with the mechanism in the shed as the water in the lock is even with the water in the level above the lock.

Lock 6 (Magazine Lock). One of the reasons the canal was not as efficient for some cargo as the railroad was that it became necessary to shut it down in the winter. Exactly when the canal closed and when it opened varied with the kind of winter in a given year. Canallers tried to get "one more trip" in, naturally, but the penalty for cutting it too close meant the possibility of being stuck in some inconvenient place for the winter. Is it as cold as it looks here at Lock 6? Yes, as the author and his family lived here for three years, we can say that it is.

Lock 7 was a popular place because of its nearness to the National Chataqua at Glen Echo which finally failed, in part due to an outbreak of malaria. The telephone lines running along the towpath were a part of a telephone system established in 1879, the longest (200 mile) existing at the time.

The Cabin John Bridge, built for vehicular traffic and to carry water to Washington from Great Falls, was begun in 1857 and completed in 1864, at the time of completion the largest single arch span in the western hemisphere. Red sandstone from Seneca and granite from Quincy, Massachusetts were carried to the site via the canal which had a lock to pass boats from the canal to a pool formed in dammed up Cabin John Creek, as shown in the foreground. (Library of Congress, c. 1864)

Lock 8, first of Seven Locks. The boat is probably the canal company pay boat. The house was constructed in 1829 and first occupied in 1830. Solomon Drew was the lock keeper at a salary of $100 a year. The house is an 18′ by 30′ stone structure with two rooms on each floor, typical of the lower part of the canal. The hut at the head of the lock is the "wait" house or "dog house" which sheltered the lock keeper at busy times when it was inconvenient to go in and out of the lock house. (National Park Service)

As Lock 9 is very close to Lock 10, the same lock keeper initially served both locks, but as traffic increased, the appearance of the frame structure, much larger than the ordinary "wait house" (which also appears here) indicates that at some time in the operating period a lock keeper was assigned to the lock. (National Park Service)

Rock (or Rocky) Run culvert between Locks 10 and 11, is one of approximately 180 culverts which carried streams under the canal, this one being more elaborate and of heavier construction than most. The culverts are outstanding examples of stone masonry seldom seen by the average person using the towpath.

Lock 11, Seven Locks. Though the stone lock houses on the lower part of the canal were almost identical, the only differences being whether the house had a central chimney or two end chimneys (which determined where the windows were on the sides of the house as well), some had side porches, or back porches or a stoop of some kind on the front. (National Park Service, 1938)

This beautiful lock house at Lock 13 was destroyed for the construction of the Capital Beltway (I-495). Why the highway had to go over this attractive spot exactly, and why highway planners usually have their way, is beyond me. The lock now sits in a "No-Mans-Land" type of setting, with no vegetation and runoff from the highway bridge eroding the lands surrounding it. Stone for the lock was "Granite from the country . . . transported by land 4¼ miles with the exception of the coping and hollow quoins, which are from Seneca." (National Park Service, 1938)

Lock 14, last of the Seven Locks, had a frame lock house rather than the stone ones typical of the lower part of the canal. The house, fence, out buildings, balance beams and the exposed walls of the locks themselves were whitewashed so as to provide more visibility. The stream coming in at the right is excess water from the canal level above carried around the lock in a bypass flume. (Brady-Handy, Library of Congress)

The Potomac Granite Company was one of the many companies which took advantage of good access to the canal and boats by building right on the berm bank of the canal. Though coal was by far the principal cargo, building materials probably came in second. (National Park Service)

An old river channel was used at Widewater to save constructing an expensive canal section, creating a very pretty ¾ mile lake, with depths up to 40 feet. Unfortunately, the river still likes to use its old channel on occasion, as it did in 1936 before the lower 22 miles of the canal were restored and again in 1942 and 1972. Massive repair work was required to restore the canal so that it could be re-watered, as shown in this 1940 photograph.

Though there is seldom a problem with a constant supply of water when a canal is built alongside a river, there is always the problem of flooding. The 1924 flood helped put an end to the canal in that year, with damage typical to that done here at the Six Locks at Great Falls.

Lock 15 is the first of the series of locks known as Six Locks, located in an old river channel which begins at Lock 18, which necessitated the building of the stone-filled log cribs from the lock to the berm shore. The lock was rebuilt in the winter of 1871-1872. Two lock houses have existed here in the past. The first was a stone one on the towpath side of the lock which was demolished in the flood of 1889. The one shown here was built about 10 years later, but was later destroyed by a fire. (National Park Service)

Operating a lock, such as in this scene at Lock 15, Six Locks, near Great Falls, required teamwork. In this case it was a family effort, with father and son manning the balance beams while mother managed the tiller. As she is wearing an apron, she might have taken time from her cooking to lend a hand in going through the lock. (Thompson Collection, National Park Service)

As at Lock 15, stone-filled log cribs are necessary at Lock 16 to contain the pool of water on the berm side of the lock, giving the lock a more massive appearance than most. It appears that the house boat (complete with portable outdoor toilet) is being towed by the steam launch in front. (Thompson Collection, National Park Service)

The 2½ story lock house at Lock 16 (Six Locks) was built c. 1837. Features of its design show that it was not built by the original lock house specifications adopted in 1828, but probably followed the specifications adopted in 1836. In spite of its closeness to Washington and the Great Falls Tavern, the family which lived here were pretty much by themselves. (Thompson Collection, National Park Service)

Again in this scene in the Six Locks area of Great Falls there is a woman at the tiller, while the men carried out the more strenuous task of opening the balance beams and perhaps helping the lock keeper crank open the wicket gates in the large wooden gates. The wooden posts at the side of the lock were the snubbing boats around which lines were wound to help steady the boat against the surging of water in the lock. (Thompson Collection, National Park Service)

Just below the Great Falls of the Potomac the canal was built between cliffs on the land side and a high heavy slope wall as high as 56 feet, with a sheer drop to an arm of the river in the deep curving Mather Gorge below. It must have been quite an experience riding along here in a canal boat! The stop gate in the background across the canal was there to stop off water from the river in time of flood.

In 1936 the house at Lock 18 was just a shell, having burned just a few years before. Today the only thing remaining is a corner of the building and some of the foundation stones.

There are many photographs of the Great Falls Tavern or Crommelin House, as it was officially known, but this one of 1889 is my favorite, showing the staff and inhabitants or habitues. The central portion (where the people are) was the original structure, built in 1829, and used as a lock house, with the enlarged portion finished in 1832. The building today is used as an Interpretive Center and as the Ranger Station for the Palisade District. (D.C. Public Library)

And what could be more natural when visiting Great Falls, but to see the falls at a fare most could afford. This old bridge went across an arm of the river at a point just opposite Lock 20. The only thing remaining of the bridge is a bit of one of the stone-filled timbered cribs. The timbers are said to have come from the Maryland Gold Mine. (Thompson Collection, Nat'l. Pk. Ser.)

Lock 22 (Pennyfield Lock) was a favorite base for President Cleveland's bass fishing expeditions and he stayed often at the Pennyfield's white frame house back on the berm. (Thompson Collection, National Park Service)

Farmers in the Seneca area used the canal to ship produce and crops to the Washington market, as attested by the storage buildings on the berm here at Riley's Lock (Lock 24). As a boat passed through the lock it went immediately into the aqueduct which carried the canal over Seneca Creek.

Boats on the canal, as cars on the highway, tend to bunch up. On the canal, boats going the same two-or-three mile per hour speed, tended to do this most often at the locks, such as these boats approaching Edwards Ferry. The boats are riding high in the water, so that we would know they were going empty or ''light'' upstream, even if we couldn't see the towpath on the left. (Thompson Collection, National Park Service)

Goose Creek River Lock. This interesting two-lock combine, completed 1837-1838, was devised to serve boats from the Goose Creek and Little River Navigation in Virginia. This river lock, the one opposite the Shenandoah River at Harpers Ferry and the one opposite Shepherdstown, West Virginia, were demanded by the Virginia legislature of 1833 to provide a market for Virginia products in return for the purchase of C. & O. Canal stock. (National Park Service, 1956)

Lock 25 (Edwards Ferry) was a busy little settlement with a granary, and general store and a ferry. The swing (or pivot) bridge which carried wagons and other vehicles across the lock to the ferry is in the closed position, where it would remain to allow passage until a boat would lock through, at which time it was swung aside. The log lying alongside the towpath was to keep towlines from snagging on the stone wall. (Thompson Collection, National Park Service)

Lock 26 (Woods Lock), is in a remote rural setting. The lock house is another one of the many which through the years has mysteriously burned, as do many other abandoned buildings in the country. The stone for the lock was boated up from Seneca. (National Park Service, c. 1940)

Broad Run Trunk is one of the most interesting features of the canal – the only wooden aqueduct carrying the canal over Broad Run, referred to as a culvert in the canal records. (Historic American Building Survey, 1961)

Granaries such as this one at Whites Ferry were common along the canal, particularly at settlements and where roads ran to the canal and river. Canal boats tied up to the wall in front of the granary where grain was discharged directly into boats by chutes. (Historic American Building Survey, 1961)

This is the view we saw several years ago of the Monocacy Aqueduct until Agnes came along in 1972 and ripped off the coping (cap) stones and much of the railing. Most of the damage has now been repaired.

This photo of Catoctin Creek Aqueduct was taken the day after its collapse on 31 October 1973. The structure had been in a weakened condition for many years prior to this date, when a local flood finished off two arches of the aqueduct, leaving only one standing. Today a foot bridge upstream of the aqueduct permits passage around it. (Jack Rottier, National Park Service)

Lock 30 (Brunswick) was one of three locks extended upstream in 1875 so as to permit two boats to pass through a lock simultaneously. When both locks were in operation there would have been a set of lock gates at the upper end of the extension, near where the post is sticking out of the ground in the photo. Brunswick has always been a busy railroad town, and the sight of railroad cars as shown here would have been a common sight to canallers. (National Park Service)

Floods have always been a problem in the Harpers Ferry area, but the 1889 flood solved the housing problem for at least one person here, below Maryland Heights. Actually, derelict canal boats provided homes for many people both during and after the canal operating period. (Thompson Collection, National Park Service)

This scene at Sandy Hook was a favorite one, the photo being used in B. & O. Railroad stations for many years, probably to deride the slowness of its rival, the C. & O. Canal. An agreement reached in the 1830's required the railroad to build a high board fence between the railroad and canal so that locomotives would not frighten the tow animals. The fence was never built, but for a time steam locomotives were unhooked at Point of Rocks and the railroad cars pulled by horses to Harpers Ferry.

It was near this point where the Shenandoah River Lock entered the Potomac River from the C. & O. Canal so as to receive boats from the Shenandoah Navigation just across the river at Harpers Ferry where the Shenandoah enters the Potomac River, as shown in the left of the photograph. Note how denuded Harpers Ferry had become through the years with the extensive use of wood for fuel and charcoal. (John Frye)

Lock 33 (Harpers Ferry Lock) about 1876 was a center of activity. All the buildings have been destroyed by flood and fire, and only the ruin of the Salty Dog Saloon remains to remind us of the events which took place here. John Cook, an advance man for John Brown, married a local girl and tended the lock here as part of his cover.

Little is known of the history of the Salty Dog Tavern which nestled against the cliff back of Lock 33. It is thought that the name is not as old as the building, as it (the name) was probably taken from a song sung in the tavern which was popular in the early 1900s. The goat in the left of the drawing may have been one of the ancestors of the wild goats which now inhabit the cliff of Maryland Heights.

There has been a dam across the Potomac River above Harpers Ferry for nearly 175 years, first for the Potomac Company for the canal around Houses Falls and for the Government Armory, later for C. & O. Canal, and now again for diverting water to the Government Canal to serve Harpers Ferry hydroelectric plant. In 1938 we could see this view of the stone-filled log and timbered cribs which formed the dam, its zigzagging course taking advantage of the rock outcrops in the river. (Nat'l. Pk. Ser.)

The C. & O. Canal was watered by seven feeder dams which impounded water to be fed into the canal in controlled amounts. The water impounded by Dam 3 above Harpers Ferry was carried to the canal below Lock 35 by this feeder canal. In the background can be seen the "mule bridge" which carried the towpath over the feeder canal. Boats were able to cross to the West Virginia shore at this point to load cargo which would then be transported down the C. & O. Canal. (National Park Service)

Another one of the industries located on the canal was that of producing lime. At one time the Potomac Refining Company had lime kilns here and operated the crusher with steam power generated by coal brought down the canal by freight boats. There was originally a manganese quarry, but the operation ceased in 1912 when the formation ran under the canal. (Nat'l. Pk. Ser.)

Antietam Creek Aqueduct is No. 4 in a series of 11 stone aqueducts on the canal. Much of this beautiful 140-foot structure was blasted down above the arches during the Civil War in 1864. (Historic American Building Survey, 1961)

Shepherdstown River Lock was one of three lift locks designed specifically to allow canal boats access to the Virginia shore and navigations. The lock was used until 1889, at which time a flood swept away the dam below here which impounded water so that boats could enter the lock. (Brooks Lucas)

The level below Shepherdstown was a favorite place to spend the night as can be seen in this canalside scene. The portable feed troughs for the mules were held in place by stakes driven in the ground. In the background is the Norfolk and Western Railroad trestle as well as the piers of the earlier Shenandoah Valley Railroad. Behind last boat is Shepherdstown Lock. (John Frye)

Bridgeport, the settlement surrounding Lock 38 across the river from Shepherdstown, West Virginia, was once a thriving area, with an inn, a feed store, and with the manse of Ferry Hill Plantation overlooking the canal. Henry Kyd Douglas, Jackson's youngest staff officer, lived at Ferry Hill and in 1859 inadvertently helped John Brown (known as Isaac Smith) climb the hill with an overloaded wagon. The bridge across the river washed out in the 1936 flood. (Brooks Lucas)

The lock house at Lock 38 (Shepherdstown Lock) was a welcome one to canallers in the early 1900's who stopped here to buy bread and pie from Cora Hebb, the wife of the lock keeper. Unfortunately, the lock house and all the out buildings were swept away in the 1936 flood. The modern James Rumsey bridge now passes directly over the lock.

Caves are one of the interesting natural attractions in the limestone area of the canal. Killings (or Killiansburg) Cave is one of the shelter caves used by the residents of Sharpsburg during the Battle of Antietam.

The Dam 4 area always sustained heavy damage during floods of the Potomac River. To minimize the effect on the canal, a stop gate was built opposite to the dam. When flood water threatened the canal, the gate was lowered via the mechanism housed in the building, thus turning aside the main thrust of the rushing water.

This canal boat scene at Lock 42 shows us several small details about canal operation and canal life. We can see a portion of the windlass by which some of the strain of opening the heavy lock gates was avoided. Fishing in the river or canal (or both) may have been good here from the nets above the out building on the right. Most children like to have pets, and these boys seem to have a pigeon in a box for theirs. Wooden biscuit boxes seem to have been handy aboard a boat. The metal objects next to the boys which at first glance seem to be dippers on closer inspection seem to be bells. It appears that they have brought out prized possessions for the benefit of the photographer.

Lock 44 in Williamsport is in an attractive setting. The frame house shown here in the post-operating period was the section or division superintendent's house. There is reason to believe that there was a lock house here at one time on the berm side of the lock. Even today the lock stones retain their nice contrasting lighter and darker colors. (National Park Service, 1961)

This vertical lift bridge over the canal built in 1923 to carry Western Maryland Railway spur over the canal to bring coal into a steam generating station at Williamsport may have been little used, or perhaps not at all, as the canal went out of operation several months after the completion of the bridge. The structure is unique and may be the smallest vertical lift bridge in the world. (National Park Service c. 1965)

Williamsport has probably always been THE canal town of the C. & O. Canal, not only from its mid-canal location, but because so many of the activities of the town were canal-oriented. Even before the canal was built, Williamsport had a lively commerce developed through the use of small boats and rafts using the Potomac River. There was always some activity here at Cushwas Wharf, where coal and other cargo was offloaded for trans-shipment. (National Park Service)

The Conococheague Aqueduct is the 5th of 11 stone aqueducts, with three equal arch spans of 60 feet each, extending 210 feet between abutments. The upstream wall of the aqueduct collapsed in 1920 when Captain Frank Myers of Boat No. 73 struck the berm wall a light blow. Capt. Myers saw the wall beginning to waver and called to his step-son, Joseph Davis, who was at the western end of the aqueduct to cut the mules loose, which he did, with no loss of life.

The Guard Lock at Dam 5 is a massive structure built in a rocky setting. From this lock canal boats entered the "Little Slackwater" behind Dam 5. In order to tow the boats from a path along the river, the mules crossed over to the land side of the lock and thence along the river to Lock 46 where the towpath once again crossed over to the river side of the canal and remained there to Cumberland. The couple standing at the lower lock gates are Lock Keeper Dan and Mrs. Sterling.

There was rather a unique arrangement above Dam 5. Unloaded or "light" boats went up the channel next to the river, but loaded boats (the ones going downstream) used the river outside the strip of land shown in the center of the photo. When the boat was not under control, there was bound to be trouble, as on 1 May 1903 at this spot when the tow boy cut or otherwise caused the mules to break the tow rope, Captain Kime and his daughter were swept over the dam. Captain Kime survived, but his daughter was killed. The frightened tow boy ran away. (National Park Service)

There are several places where it is necessary for mules to cross from one side of the canal to the other. One of the places is at Lock 46 (Two Locks) where the mule crossover bridge carries the towpath from its short stint on the land side of the canal from Lock 45 back over to its normal position on the river side of the canal.

Charles Mill below Four Locks. There are three Charles Mills on the canal, or which have had the name Charles Mill sometime in their history. Most mills changed names as the owners changed, but as far as we know, the mill here has always been Charles Mill. The mill operated an 18½′ diameter by 4′6″ wide overshot water wheel to run this old grist mill, which at one time was a plaster mill according to historical records. The mill took water from Camp Run, which was its normal source of water power, and from the canal in periods of low water. (National Park Service, 1916)

Four Locks. The four closely spaced locks here were constructed to raise the canal bed some 32 additional feet in order to cross Prather's Neck. The river makes a five mile long loop at the point, and the construction of these locks saved about four miles of canal construction. To the extreme right was a drydock on the berm side of Lock 47 and this photo shows a boat in the dock. Note that the building at the next lock upstream, Lock 48, was located over the bypass flume and used the water of the flume to man machinery in the shop inside. (John Frye)

The old mule barn above Lock 50 at Four Locks was used to keep feed and to house mules over the winter period when the weather closed the canal. It is of frame and typical early period mortice, tenon and wooden pin frame construction, and seems quite old except for the rafters, nailing strips and the galvanized iron roof. The barn was restored by the Park Service in 1975. (National Park Service c. 1960)

One difficulty about a canal is getting from one side to the other. With the lock gates closed, one could clamber across the catwalk, but that wasn't always so easy. On the other hand, bridges were a problem as they interfered with the passing boats. Here at Fort Frederick are two of the ways that the crossing problem was handled; a permanent bridge, 17-feet high was built for pedestrians, and a swinging or pivot bridge was built for vehicles. The bridges were placed at this narrow structure formed by the masonry for the stop gate, placed at the bottom of Big Pool so as to prevent the escape of water when repairs necessitated draining the level below. (National Park Service)

Hancock, Maryland is one of the principal towns along the canal, an important road center even before the construction of the canal. Old names for the town were The Connolloways, North Bend and Hancock Town. Old taverns in the town (no longer existing) were Sign of the Cross Keys, Sign of the Green Tree, Sign of the Ship, Union Inn and Sign of the Seven Stars. The old bridge over the canal and Potomac River seen in the background washed away in the 1936 flood.

This is the Round Top Cement Company above Hancock in 1907. In point of fact, the cement plant began operations in 1837, and much of the cement and mortar used in the masonry of canal structures was produced here. First known as Shafer's Cement Mill, it changed to the present name in 1863. The shed-like structure on the right is in front of the row of eight lime kilns; the smoke stack on the left is part of the steam generating plant, used to power the machinery for grinding lime burned in the kilns. (Mr. Truax)

This was a busy area for many years, for it was here at Dam 6 that the canal terminated from 1842 until 1850, with canal boats passing to the West Virginia side of the river to pick up coal from the railroad, probably through this feeder canal. The boat is passing through Lock 53 and will soon be passing through Lock 54 just ahead. Unfortunately, the lock house and the out buildings are now all gone. (National Park Service)

This tranquil scene belies the fact that floods never failed to leave their mark on Dam 6. The 1877 flood seriously damaged the abutments; the 1886 flood tore a hole in the center of the dam; the canal-killer, the "Johnstown" flood of 1889 virtually destroyed the structure, completely washing away 100 feet of the breast, but it was back in service in 1891. The 1924 flood which helped to finish off the canal did little damage, but the 1936 and 1942 floods turned the structure into the rubble one sees today. This may be the only dam which burned, as the 475 foot wooden structure of the dam was accidentally set afire by a fisherman and blazed away out of control.

This 1938 photograph of Dam 6 shows the detail of the dam which we could see at that time. Today there are only a few of the timbers left, though there is considerable rock rubble.

The frame board-and-batten lock house still exists at Lock 56 (Pearre). There used to be a general store near the lock and an old hotel nearby, but both are now gone. The lock was completed in 1848 and a George Murkey was once a lock keeper here. (Historic American Building Survey, 1961)

Lock 58 is located in a remote part of the canal. Though it doesn't appear so, this was the first of 13 locks to use roughly-cut and coursed stone in the lock walls covered with treated wood. Lack of suitable building material and a paucity of funds were the reasons for going to this cheaper treatment. The treatment of the bypass flume in these upper locks is more formal than those farther down, with a wall continuing into the berm bank. The frame lock house is no longer standing. (National Park Service)

Locking through, such as this boat is doing at Lock 60, gives those on board a chance to stretch their legs, and those on shore something to look at. Undoubtedly, the ones most appreciative of the break in the routine are the mules which appear to be taking their leisure in this scene. Another day will put this light boat back in Cumberland for another load of coal.

Busey Cabin is a canal structure seldom seen by canal travelers, either today or in the past, as it is located up Robey Hollow in a very remote part of the canal off Kasecamp Road. The cabin is constructed of hewn logs chinked with mortar and was used as a construction office during the building of the canal. (Historic American Building Survey, 1961)

The caption of this old photograph said that this is "At the lock below the cut leading to the Canal Tunnel," which should make it Lock 62 from the terrain and the elevation of the lock at the upper end. The old log lock house has long since disappeared in this lonely part of the canal. (The Smithsonian Institution)

At several places along the canal were shops owned by the canal company for maintenance, repair and the manufacture of canal items. This building at Lock 66 was a Carpenter Shop, here lock gates were made. At the time the photo was taken in about 1950, there was still a lock gate being worked on when the building was abandoned. The shop is now gone, except for some rubble, as is the lock house and all out buildings.

The interior of Lock 66 shows the type of wooden sheathing used over the rough stones used in the lock walls. A row of vertical studs was first attached to the stone, and then a layer of horizontal planking followed by a vertical layer of planking. Concrete used in 1906 and after, at this lock in the gate recesses. (National Park Service, 1958)

It has always been an adventure to go through the 3,118-foot Paw Paw Tunnel, whether on a working boat, an excursion boat such as the OAK SPRING, or by foot. Work was undertaken in the tunnel from 1834 until 1850, though it was holed through in 1840. The tunnel cut leading to the downstream end of the tunnel necessitated the use of a wooden towpath for some distance. Much of this end of the tunnel is now filled with debris from a landslide, a frequent occurrence here.

The enormity of the task of constructing the Paw Paw Tunnel can be appreciated partially by noting the tremendous cut needed to even approach the tunnel. At the downstream (or north) portal the tunnel is lined first with stone, and then with brick four courses thick except under the vertical ventilation shafts where it is six. The tunnel is 24′ wide and has 11-foot sidewalls. The towpath runs on a ledge about 4′. The height of the tunnel is 24½′, 17½′ above the water line. (Photograph is in the 1960's)

As much as I have studied this old photograph, I cannot identify it. George W. "Hooper" Wolfe, early towboy and author of I DROVE MULES, says that it could be one of the Oldtown Locks. The setting is similar to Lock 70, but the house is different than that existing there today. Do we have any other guesses? We do know that George Swandol was the captain of Boat No. 10 in the 1918-1924 period. (National Park Service)

Railroads cross the canal many times between Georgetown and Cumberland, this crossing being Bridge No. 65 of the canal's rival, the Baltimore and Ohio Railroad. Railroaders would rile canallers by shouting "Muskrat on the tiller!" and similar phrases, and the canallers would have similar choice words in return. The "Three Locks" at North Branch are once again built of stone. Lock 74 is in the foreground and Lock 75 is just ahead. (National Park Service)

Lock 75, the last of "Three Locks" at North Branch, is the last lift lock on the canal. At the lower end of the lock near the man, who is probably the lock keeper, we see a type of roller windlass used to crank open the lock gates. Just below this we see a stone-filled log crib which diverted the rushing water of the bypass flume from the side of the boat entering or leaving the lock. The lock house is log chinked with mortar, with a board and batten wooden addition. (National Park Service)

Mules were very essential to the operation of the canal. Though horses were used in the construction of the canal, we see few references to them in later years and hardly a photograph has a horse in it. The scene here at Lock 75 at North Branch shows very nicely the harnessing arrangement and the attire of the tow boy, also called a ''hoggee'' and sometimes a ''mule skinner''. (National Park Service)

Much time was spent by the lock keeper in his ''shack'' or ''shanty'' or ''dog house'' or ''wait house'' or any one of the number of names given to the small structure which gave him shelter. Normally this building was located at the upstream end of lock (for visibility) and, at this Lock 74, to give a little protection to the drop gate hoisting machinery and to the operator. It was located on the berm side of the lock so as not to interfere with the operation of the towline. Note the improvised seat, a flag pole, the smoke stack for his stove, a lantern, a wooden bucket for water, etc. To the right is the pivot bridge which crossed this lock. (National Park Service)

Cumberland, Maryland is the place where the loading of coal took place, the place where canallers felt at home and lived and caroused and wintered. The Cumberland end of the canal was a complex place with basins and wharfs and loading platforms. Today the area is filled in and it is difficult to get the feeling of the activity which took place here during the busy operating days of the canal.

Cumberland, Maryland. The loading of coal took place directly from railroad cars into the boats below. Either one or two boats are being loaded here and the others are awaiting their turns. The horse, used to pull the railroad car, is a marked contrast to all the mules we have seen on the canal. C & P on the railroad cars stands for Cumberland and Pennsylvania, whereas the C. C. C. stands for the Consolidation Coal Company. (Thompson Collection, National Park Service)

The twin Cumberland guard locks at mile 184.5 mark the end of the canal. A lock keeper's house once sat here. Skat Eaton and his family long lived in the building here between the two guard locks, until about 1957. The railroad, once the canal's rival, dominates the scene by passing heavily overhead the lock, seemingly oblivious of its once important role. (National Park Service, in the 1950's)

This post-operating period (after 1924) photo shows the terminus at the guard locks into the Potomac River. The Western Maryland bridge passes over Dam 8 which impounded water so that canal boats could operate above here, but more importantly, so that water would flow into the canal, as the water flowing in here would supply 50½ miles of canal – down to Dam 6. The openings of the guard locks have now been filled in with concrete. (National Park Service)

The destruction of Dam 8 and the infilling of the canal basins in Cumberland marked the end of the integrity of the old canal in Cumberland. Cumberland now seeks recognition for the end of the canal it once had and would seek again. The Corps of Engineers destroyed this dam as part of a flood control project. It seems odd that the engineers always seem to want to build dams in rivers where they don't exist, but if there is an old historic dam, it has to go. (National Park Service)

Other Canal Publications Offered by the American Canal and Transportation Center

GEORGE WASHINGTON'S CANAL AT GREAT FALLS, VIRGINIA – A history of the Potomac Company and other ventures which attempted to make the Potomac River navigable plus a Towpath Guide to the Potomac Canal at Great Falls Park. About 45 maps, photos, drawings. 48 pp. $2.50

TOWPATH GUIDE TO THE C & O CANAL, SECTION ONE, Georgetown to Seneca – (Revised 1974) 56 pages, 7 maps, 33 photos. $3.00

TOWPATH GUIDE TO THE C & O CANAL, SECTION TWO, Seneca to Harper's Ferry – 52 pages, 8 maps, 28 photos. $3.00

TOWPATH GUIDE TO THE C & O CANAL, SECTION THREE, Harper's Ferry to Fort Frederick – (Revised) 64 pages, 6 maps, 33 photos. $3.00

TOWPATH GUIDE TO THE C & O CANAL, SECTION FOUR, Fort Frederick to Cumberland – 68 pages, 10 maps, 35 photos. $3.00

(**$10.00** for set of Four Towpath Guides)

LIFE ON THE CHESAPEAKE AND OHIO CANAL (1859) – By an itinerant anonymous New England peddler. Many historic photos. $2.50

I DROVE MULES ON THE C & O CANAL – Fourth and last edition of this colorful hardbound 210 pp. book, 8 x 10 in., contains 100 stories of rough-and-ready life on the Chesapeake and Ohio Canal by mule driver George W. "Hooper" Wolfe, 120 photos. $15.00

FERRY HILL PLANTATION JOURNAL – 1975 Reprint of Life on the Potomac River and C & O Canal, Jan. 4, 1838 to Jan. 15, 1839 – 170 pages, with illustrations. $4.95

MITCHELL'S COMPENDIUM OF CANALS AND RAILROADS IN USA, 1835 – A faithful facsimile of the original, 84-page, 3¾ x 6¼ in. just as published by Mitchell & Hinman in the early 1800's. $1.95

THE PENNSYLVANIA MAIN LINE CANAL – 1976 Reprint of this colorful account of the early canal-railroad system connecting Philadelphia and Pittsburgh, by canal researchers McCullough and Leuba. 218 pages, fully illustrated. $4.75

THE AMAZING PENNSYLVANIA CANALS – 1975 Reprint of this definitive work on all of Pennsylvania's historic 1200-mile canal system, by Engineer-Historian William Shank. 80 pages, 68 illustrations, some in full color. $3.50

CANAL DAYS IN AMERICA – A 311 page, hardbound, coffee-table size book on the historic American Canals by Harry S. Drago. $10.00

MAP OF THE CANALS AND RIVERS OF THE UNITED STATES – (1905) $2.00

THREE HUNDRED YEARS WITH THE PENNSYLVANIA TRAVELER – 1976 First Edition of this 8½ x 11 in., 200 page, two-color picture history of all forms of travel in the Keystone State since the time of the Indians. Text by William H. Shank, P.E. $13.50

C & O CANAL COMMEMORATIVE MEDAL – celebrating the 150th anniversary of the beginning of the canal. Bronze, 39 mm (1½ in.) diameter. Lock scene on one side, company seal on the other, rigid case. $4.50

(Orders may be addressed to American Canal and Transportation Center (WV), Box 842, Shepherdstown, WV 25443. Add 50¢ for shipping. In West Virginia add 3% to cover State Sales Tax. Send a self-addressed stamped envelope for a complete listing of our publications.)